ISBN 978-0-666-74973-4
PIBN 10858001

THE SUBJECT MATTER OF A COURSE OF SIX LECTURES ON THE NON-METALLIC ELEMENTS.

BY

PROFESSOR FARADAY.

DELIVERED BEFORE THE MEMBERS OF THE ROYAL INSTITUTION, IN THE SPRING AND SUMMER OF 1852.

ARRANGED, BY PERMISSION, FROM THE LECTURER'S NOTES, LENT FOR THE OCCASION

BY

J. SCOFFERN, M.B.,
Late Professor of Chemistry at the Aldersgate College of Medicine

TO WHICH IS APPENDED,

REMARKS ON THE QUALITY AND TENDENCIES OF CHEMICAL PHILOSOPHY, ON ALLOTROPISM, AND OZONE; TOGETHER WITH MANIPULATIVE DETAILS RELATING TO THE PERFORMANCE OF EXPERIMENTS INDICATED BY PROFESSOR FARADAY.

LONDON:
LONGMAN, BROWN, GREEN, AND LONGMANS.
MDCCCLIII.

LONDON: PETTER, DUFF, AND CO. PRINTERS,

PLAYHOUSE-YARD, BLACKFRIARS.

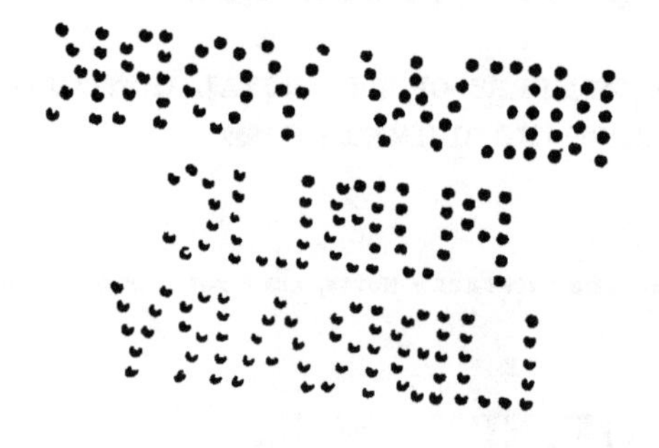

TO

WILLIAM THOMAS BRANDE, ESQ.

OF THE ROYAL MINT,

F.R.S., M.R.I., &c. &c.

Professor of Chemistry at the Royal Institution,

THE FOLLOWING PAGES ARE INSCRIBED, AS A SLIGHT MARK OF RESPECT AND ESTEEM,

BY

J. SCOFFERN.

Professor Brande's Lectures on the application of Organic Chemistry to the Industrial Arts, are in course of reproduction from his Lecturing Notes, and, with his sanction, they will shortly appear.

PREFACE.

HAD this volume consisted of Professor Faraday's Lectures alone, it might have gone into the literary world without further preface than such as is conveyed in the intimation of its appearance with the fullest sanction and consent of the lecturer. An explanation, however, is rendered necessary, when, as in the present case, an editor intersperses additions of his own.

It may suffice on this topic to state that the extraneous portions of the volume suggested themselves during an interview with Professor Faraday, in the course of which the process of rendering an oral discourse into a literary shape formed the topic of conversation. It was conceded that lectures, for the most part, have reference to others already delivered; that a lecturer frequently indicates collateral facts merely; not demonstrating the line of evidence by which these facts had been arrived at—because so far as might relate to a particular audience the knowledge of such facts was assumed. It was conceded, moreover, that a chemical lecturer, more perhaps than any other, possessed a means of demonstrating facts not available to the essayist—*the demonstration of experiment*—that mute eloquence of action which silently compresses whole pages of written lore into one short act of manipulation, and renders verbal explanation unnecessary.

These and many other special peculiarities, in which the functions of a lecturer differed from those of an author, having been discussed—it was conceded that a mere verbatim report of an experimental course of lectures, would by no means render, under a literary aspect, the spirit in which these lectures were delivered.

Accordingly—being anxious to obtain these lectures for a public journal, it appeared that the object would be most efficiently secured by attending them regularly—embracing their scope, noting their experiments, striving to imbibe their philosophy—and transferring their language to paper, whenever language, not experiment, might be the form adopted for expressing a sentiment, or inculcating a truth.

This plan was adopted accordingly: and to render it still more perfect, Professor Faraday kindly and cordially furnished, immediately on the conclusion of each discourse, his lecturing notes: moreover, whenever any difficulty occurred, he no less kindly lent the aid of his supervision. Originally intended for the pages of a journal, the rendering of Professor Faraday's lectures was necessarily much condensed; when therefore public appreciation had made a fuller expansion of them desirable, the lecturing notes of Professor Faraday proved of redoubled utility; containing, as they did, various memoranda of points indicated for discussion, but not touched upon during the lecture for want of the necessary time. These dormant notes I have frequently taken the liberty to expand.

Enough will have been stated to make known the warrantry under which I have acted in rendering the lectures themselves; and it equally applies as accounting for the existence of those parts of the volume which are my own: parts

ful metamorphoses which it discloses, by the protean display of physical changes which it brings before our eyes, by the demonstration it affords of the indestructibility of matter under the agency of existing laws, is perhaps more calculated than any other science to awaken within us the most ennobling sentiment the mind can contemplate—the sentiment of immortality. If the grosser parts of our earth and its inhabitants pass thus undestroyed through all the vicissitudes of death, fire, and decay, how impossible is it to assume a destruction of a spiritual essence! Totally irreconcilable with the genius of chemical science is the idea of destruction.

Chemistry is essentially a science of experiment;—most of the conditions under which its phenomena are developed requiring the disposing agency of man: yet there occur naturally a sufficient number of chemical phenomena to rivet the attention of a reflective mind, and lead it to some acquaintance with that mutation of form apart from destruction which is so striking an attribute of chemistry.

Many a reflective sage must have speculated ere now during the very infancy of the world, and long before chemistry had invoked the aid of experiment, on the cause and consequence of such ever-recurring phenomena as combustion and evaporation. The circumstance must have been noticed that the waters of lakes and streams, although exhausted in vapour by the agency of the sun's rays, and dispersed, were not dispersed to be destroyed; but that, entering into clouds, the aqueous vapours remained hovering above until the operation of some natural cause should effect their descent to earth once more in hail or snow, dew or rain. Many a reflective sage in ancient times must have speculated on the results of combustion; more subtle, less amenable to scrutiny than those of evaporation though they be; and although neither the priests of Isis nor the sages of Greece knew the means of collecting, as does the chemist of our own times, the fleeting gaseous elements which are scattered by combustion, yet the spiritual intuition of these philosophers anticipated in a poetic myth the slower evidence of induction. In the fabled rising of the Phœnix from her ashes is displayed a credence in the non-destructibility of matter under the operation of existing laws; and the many changing aspects of Proteus seem but the expression of a belief in the occurrence of chemical transformations.

No sooner were the manifestations of chemical action displayed by experiment, than the wonderful mutations of form as evidenced by combination and decomposition gave rise to hopes that the baser metals might be transmuted into gold; a credence which, when we come to investigate the period of its first origin, carries us back into the furthest recesses of antiquity.

Viewed under the aspect of a prevailing chemical furor, the belief in alchemy may be said to belong to the middle ages; but amongst all its votaries a sort of universal credence was given to the remote origin of the doctrine. The Egyptian Hermes* was held to be the accredited originator of alchemy, and the notion of the possibility of transmuting base metals into gold has been thought by some writers to have existed amongst the Greeks at a period coeval with the Argonautic voyage in quest of the Golden Fleece. Suidas, in his lexicon published during the 11th century, expressly states under the head "Δερας," that the golden fleece was only a mythical expression for a parchment document on which had been written a description of the process for making gold.

Without minutely discussing the opinions or the history of alchemical writers, it may suffice to remark that a belief in the possibility of making gold, and extending the duration of man's corporeal life, constituted the foundation of alchemy. Not that these tenets were received by all professing alchemy to the same extent; in this respect the greatest difference of opinion obtained. Some votaries merely contented themselves by expressing their belief in the possibility of transmuting the base metals into gold, whilst others proclaimed to the world in mystic terms their possession of the secret.

Rescued from the obscure jargon of the language in which these descriptions were veiled, the opinions of

* Hermes Trismegistus.

the alchemists as relates to the composition of metals, were far from irrational,—always remembering the kind and the amount of information at their disposal. All the metals, they believed to be compounds:—the baser metals containing the same elements as gold, from which they differed on account of their association with impurities. These impurities being separated, it was imagined that gold would remain. The agent supposed to be capable of effecting this purification was the philosopher's stone, which, although many alchemists did not hesitate to state they had made, the greater number limited themselves to the expression of a belief in its existence.

It would be ungrateful in the chemical philosopher of the present day to contemn utterly the striving of the alchemists. Many of these enthusiasts there were who shadowed forth in their hyperbolic phrases the sterner facts of induction; and even the labours of that sordid class who had no nobler aim in view than the amassing of wealth, disclosed a vast store of collateral facts for the benefit of future chemistry. Nor is it, perhaps, just to stigmatise by so harsh a term as insanity that belief in an elixir which should be capable of extending the life of man, in his corporeal form, beyond the limits allotted to mortality. Having started from the basis of considering gold a noble metal, untarnishable in the air, imperishable in the fire, unalterable by all common solvents, it does not seem a flight of imagination beyond the confines of sanity although fanciful, to assume that the human frame might be so imbued with gold as to be proof against many of the destructive agencies to which it is ordinarily subjected. Many a recent medical hypothesis has rested on a basis far less seemingly rational than this.

The hope expressed by many alchemists of their being enabled to extend the life of man beyond that of the patriarchs—nay, even to render him immortal in his corporeal form—was a wild flight of fancy starting from a basis of seeming probability; a pushing to extremes of a theory in itself not so much at variance with given premises as many have conceived. Doctrines in themselves rational are often strained by ardent minds until they assume the semblance of error,—theories often forced beyond their proper sphere,—until, breaking loose from restraint, they lead where they ought to follow, suggesting accordances rather than associating facts. Such tendencies are common to all doctrines in all times. It is a quality of the human mind to be ever striving at perfection, ever aiming at the acquisition of that which seems to be true. The isolated fragments of truth which lie scattered in our path, we are ever endeavouring to bring together or arrange. Too confident in the strength of our own perceptions—too oblivious of the narrow limits which restrain the excursions of our reasoning, we are ever prone to set in order, and construct into a fondly-thought temple of perfection, those scattered fragments of truth. The temple stands approved by our own complacent scrutiny, —we die, and others fill our place. Then comes the inductive reasoning of a future age, and proves the cherished edifice of truths so pleasing to our eye, to be a monstrous distortion.

During the period of a century or more, it was the custom to spurn the doctrines of the alchemists; not only in the literal acceptation of these doctrines, but even as semblances of philosophic truths. The time has passed for this opinion to be maintained. Within the last few years a series of manifestations has been noticed which goes far to vindicate many opinions of the alchemists. The condition of *allotropism*, or the quality which certain bodies possess of assuming two marked phases of chemical and physical existence, shatters the opinion on which our absolute repudiation of the doctrine of transmutation was based. Chemists now regard the idea of transmutation, not so much in the sense of being absolutely, essentially false, as a vision of truth distorted by the aberration of the medium through which it has been made to pass. Although a belief in the existence of an universal elixir and the philosopher's stone was not extinct even so late as the beginning of the present century,* yet the deca-

* I allude to Peter Woulfe, who died in the year 1805, of whom remarks Professor Brande (Manual of Chemistry), "it is "to be regretted that no biographical memoir has been pre- "served. I have picked up a few anecdotes respecting him from "two or three friends who were his acquaintances. He occupie

lence of alchemy as a popular infatuation, may be said to date from the time of Paracelsus. He, in despite of his wild absurdities, had the merit of strengthening the alliance between chemistry and medicine, by rendering the public familiar with the exhibition of certain medicinal preparations, and of contributing to the facts of chemistry by many original discoveries.

The life and writings of Paracelsus seem naturally

"chambers in Barnard's Inn while residing in London, and "usually spent the summer in Paris. His rooms, which were "extensive, were so filled with furnaces and apparatus that it "was difficult to reach his fireside. Dr. Babington told me "that he once put down his hat, and never could find it again, "such was the confusion of boxes, packages, and parcels that "lay about the chamber. His breakfast hour was four in the "morning: a few of his select friends were occasionally invited "to this repast, to whom a secret signal was given by which "they gained entrance, knocking a certain number of times at "the inner door of his apartment. He had long vainly searched for "the elixir, and attributed his repeated failures to the want of "due preparation by pious and charitable acts. I understand "that some of his apparatus are still extant, upon which are sup- "plications for success, and for the welfare of the adepts. When- "ever he wished to break an acquaintance, or felt himself "offended, he resented the supposed injuries by sending a present "to the offender, and never seeing him afterwards. These "presents were sometimes of a curious description, and consisted "usually of some expensive chemical product or preparation. "He had an heroic remedy for illness; when he felt himself "seriously indisposed, he took a place in the Edinburgh mail, "and, having reached that city, immediately came back in the "returning coach to London. A cold, taken on one of these "expeditions, terminated in an inflammation of the lungs, of "*which* he died in 1805."

to indicate the point of immergence of the first chemical epoch into the second. Before the time of this wild enthusiast, it had been the universal custom to refer unknown agencies to the operation of supernatural causes, a custom which, however much it may have been conducive to the free play of poetic imagination, is but little accordant with the genius of inductive philosophy. With Paracelsus this universal belief in occult agencies may be said to have ceased; and, as the expiring flame burns sometimes brighter before its final extinction, so did the belief in occult causes, before its final extinction, become intensified in the mind of Paracelsus. He was, perhaps, the most universal Pantheist the world had seen. Every existing thing, however seemingly inanimate, was invested by Paracelsus with an ideal life, and was assumed to perform its vital functions under the guidance and direction of some familiar spirit. The popular names of many chemical substances at this day bear ample testimony to the general credence which must have prevailed in the early days of chemical, or more properly speaking, alchemical philosophy. The words *spirit of wine, spirit of salt, spirit of nitre,* and so forth, sufficiently bespeak a general belief formerly existing in respect of supernatural or occult agencies.

"Paracelsus," remarks an able French author, "may be regarded from many points of view; but the one which will interest us most relates to the impulsion, the new direction which he imparted to chemistry; and although it would be difficult to collate the views of that singular individual into an ordinary scheme, yet I cannot help referring to him as the originator of various theories.

At the commencement of the sixteenth century, demonology and the cabalistic art pre-occupied the minds of all. In vain had astrology been proscribed by a Papal bull and by the faculty of Paris; in vain had alchemy been interdicted by the Senate of Venice; the pretended sciences still continued to be taught in the majority of Schools. Men of true science did not scruple to become their advocates. George Agricola, Jean Bodin, Jerome Carden, Thomas Erastes, were amongst its most strenuous advocates. Felix Plater, Ambrose Paré, nay, even the judicious Femel, did not hesitate to lend it their faith. During the same century, the plague and other epidemics broke out; to remedy them, the populace had, first of all, recourse to astrology and cabalistic practices; then they tried the manifold pharmacy of the Arabians, which, failing in its turn, recourse was at last made to alchemy, which offered a host of new remedies, some of them not devoid of efficaciousness.

In this condition was the popular mind when Paracelsus presented himself as the instigator of a radical reform in medical doctrine; but his audacity, his disdain, his violent attacks against admitted opinions, although unanswered, did not even suffice. He therefore invented new doctrines—doctrines which, he said, would overthrow all that had gone before. He commenced by emitting a sort of physiological theory, founded upon the application of astrology to the functions of the human body. He regarded magic as the culminating point of all science, and he believed its study to be a matter of prime importance to all destined for the medical profession. Finally, he went the length of affirming that, by means of the cabalistic art and chemistry, health might be re-established, life might be extended, and that even animated beings, *homunculi*, might be produced.

As regards the intro-chemical doctrine, of which he must be regarded the founder and the chief, it reduces itself into this proposition; that the composition of the body of man being formed by the union of a *sidereal* (that is to say, immaterial) salt, a sulphur, and a mercury, and diseases being caused by the alteration of this compound, they could only be combated by chemical means, combined with astral influences.

With the view of explaining the action of medicaments, he substituted for the elementary qualities of Galen, a reasoning being—the *archæus*, which seemed to discharge the part of nature in the play of our organs; which combined elements, elected the materials of nutrition, ejected impurities, and re-established the equilibrium of physiological functions. The archæus was, in the doctrine of Paracelsus, the spirit of life, the sidereal portion of man's body; hence, we may regard it as pretty nearly identical with the vital principle of the moderns. Having adopted the system of Genii, Paracelsus attached to each natural object its own familiar spirit, to which he gave the name of *Olympian* spirit. Hence arose the supposed relation between man and various natural objects; and this belief, so long admitted, it was that attributed to certain substances qualities having reference to their natural forms. It will be evident that Paracelsus was the originator of dogmatic chemistry. From this moment henceforth, a strong line of demarcation was drawn between chemists and alchemists. The latter continued their vain researches, whilst the former pursued the application of positive facts to medicine and the industrial arts.

Notwithstanding all that was vague, and singular in his ideas, it cannot be denied that Paracelsus advanced science by his own researches, and by the discovery of many facts. It was he who first offered a true chemical and medical view of the preparations of antimony, mercury, and iron. He was the first to promulgate the doctrine that certain poisons, when taken in fitting doses, act as safe and efficient medicines. He was the first to recommend the employment of preparations of lead in diseases of the skin. In like manner he employed copper, nay, even arsenic, as external applications for effecting cauterization. He employed sulphuric acid in diseases resulting from poisoning by lead, a mode of treatment which has remained in vogue up to the present day. He was the first to distinguish alum from copperas, remarking that the former contained an earth, the latter a metal. He was the first who mentioned zinc, regarding this metal, it is true, as a modification of mercury and bismuth. He admitted the existence of other elastic fluids besides atmospheric air, specially alluding to hydrochloric acid gas, and sulphurous acid gas, both of which, however, he regarded as forms of water and of fire. According to him, the spark of flint and steel was the product of fire contained in the air. He was the first to point out, that when oil of vitriol was made to act upon a metal, air was disengaged, which air was *one of the elements of water*. He was aware that the presence of air was indispensable to the respiration of animals, and the combustion of wood. He stated that calcination *killed* those metals subjected to it, and that *charcoal reduced or revived* them. "There exists a certain thing," said he, "which we do not perceive, and, in the midst of which is plunged the whole universe of living beings. This thing comes from the stars, and we may obtain a notion of it in the following manner:—Fire, in order that it may burn, requires wood, but it also requires air. The air, then, is the life, for, if air be wanting, all living beings would be suffocated, and die." In another part of his writings, Paracelsus states that digestion is a solution of aliments, and that putrefaction is a transformation: that, moreover, everything dies previous to resuscitation in some other shape. These great chemical and physiological views—this generalization of the phenomena of combustion and respiration—prove that, despite his drunken hallucinations, Paracelsus held within him a spirit of deep penetration.

theory. The illustrious Boyle, who does not seem to have been aware of the doctrine of Beccher and Stahl, although it was promulgated before his death, performed a very remarkable experiment, which, had he subjected to mature reflection, might have led to the discovery of oxygen gas, and to an overthrow of the phlogistic theory. Having fused some tin in an open glass vessel, and retained it for some time at the melting temperature, he allowed it to cool, and weighed the result, which was now found to be heavier than the original tin. This experiment is merely cited by Boyle with the view of proving the materiality of heat, and, as interpreted by him, is, in so far, inconsistent with the phlogistic theory, that, according to Boyle, the calx or oxide of tin was not tin *minus* something, *i. e.*, the imaginary phlogiston, but *plus* something,—*i. e.*, heat, or the matter of heat.*

---

* In reference to this experiment of Boyle, Professor Johnstone has made the following pertinent remark :—" How much the progress of science depends on the mode in which phenomena are interpreted by the first observers, is strikingly illustrated in the case of certain experiments of Robert Boyle. He observed that, when copper, lead, iron, and tin, were heated to redness in the air, a portion of calx was formed, and there was a constant and decided increase of weight (*Experiments to make fire and flame ponderable.* London, 1673). This experiment he repeated with lead and tin in glass vessels, hermetically sealed, and found still an increase of weight; but observed further, that when the '*sealed neck of the retort was broken off, the external air rushed in with a noise*' (*Additional experiments, No. V., and a* [illegible] *the perviousness of glass*

Very soon after the time of Stahl, the gain of weight acquired by substances after combustion had become a phenomenon so generally known, that such as advocated the phlogistic theory were obliged to devise some other explanation for the phenomenon. Accordingly, the hypothesis was assumed, that phlogiston, in itself, possessed the quality of levity; hence, that its presence must confer the property of lightness; indeed, the phlogistic theory was held with extraordinary pertinacity, until the discovery of oxygen by Priestley rendered its further acceptation untenable.

In the year 1774, Dr. Priestley, being at Paris, demonstrated to Lavoisier that oxygen gas could be extracted from red oxide of mercury, a demonstration which caused Lavoisier to become a powerful advocate

---

to ponderable parts of flame,—Exp. iii.). From this he reasoned correctly, that in calcination, the metal lost nothing by drying up, as was generally supposed, or that if it did, 'by this operation it gained more weight than it lost.'—Coroll. ii. But this increase of weight he attributed to the fixation of heat, stating it as 'plain that igneous particles were trajected through the glass,' and that 'enough of them to be manifestly ponderable did permanently adhere.' Had he weighed the sealed retort before he broke it open, he must have concluded that the metal had increased in weight at the expense of the enclosed air. He stood, in fact, at the very brink of the pneumatic chemistry of Priestley; he had in his hand the key to the great discovery of Lavoisier. How nearly were those philosophers anticipated by a whole century, and the long interregnum of phlogiston prevented! On what small oversights do great events in the history of science, as of nations, depend!"—Johnstone, Trans. Brit. Ass. vol. 7, p. 163.

of the anti-phlogistic doctrine. In 1785, Berthollet also adopted the new view. Fourcroy and Guyton de Morveau then joined the ranks, and, from this time forward, the anti-phlogistic theory continued to advance in public estimation. Its universal adoption may be said to close the second great epoch of chemical progress.

III.

IRD CHEMICAL EPOCH—DISCOVERIES AND GENERALISATION OF LAVOISIER—NEW CHEMICAL NOMENCLATURE—DEVELOPMENT OF THE LAWS OF CHEMICAL COMBINATION—THE ATOMIC THEORY—CHEMISTRY BECOMES GRADUALLY ALLIED WITH MATHEMATICAL NOTATION, AND ASSUMES THE FORM OF AN EXACT SCIENCE.

o sooner was the phlogistic theory overthrown, than avoisier and his associates began to furnish that nomenature of chemical bodies which, in all its essential rticulars, still remains, although inconsistent with any facts at this time known. The nature and outline this nomenclature are so well known as to render uncessary any remarks on the subject. The nomenclare is one of great simplicity and beauty, but affords a riking instance of the disadvantage resulting from the loption of theories as the ground-work of systematic rangements. By a too hasty generalisation of a limited umber of facts, the substance oxygen was assumed to the universal former of acids, and hence its name. his assumption of the universal acidifying quality of xygen, is a fault which lies at the basis of Lavoier's nomenclature, and prejudices its structure. The eater number of acids, it is true, contain oxygen, but any are without it—and even those which contain cannot logically be said to manifest the quality of

The proportionality of chemical composition and decomposition attracted but little notice from the time of Richter until between the years 1803* and 1808,† when Dalton, speculating on the rational consequences of the facts made known by chemical decomposition, first promulgated a theory of atoms, based upon arguments no less rational and intelligible than the theory of Epicurus was irrational and obscure:—a theory which carries with it so many elements of conviction, that a celebrated modern chemist has concluded his arguments for and against the existence of atoms by the remark, "*that whether matter be atomic or not, thus much is certain, that granting it to be atomic, it would appear as it now does.*"

The arguments adduced by Dalton in favour of the existence of atoms, were such as these:—seeing that all marked cases of chemical combination can be demonstrated always to take place in definite proportion, and that by inference, a similar proportionality may be supposed to extend to less marked cases—seeing that these definite proportions of bodies entering into combination are mutually proportional amongst themselves, it follows that such definite immutability, such mutual proportionality, should most rationally be considered as indicating a ponderable ratio between combining elements; and that the ratio never changing would seem to be indicative of elementary ponderable molecules of determinate

* Manchester Phil. Trans. for 1803.

† Dalton's New System of Chemical Philosophy.

relative weight, unchanging, indivisible; qualities which will be recognised as fulfilling the definition of *an atom*.

With the view of practically illustrating the train of reasoning by which Dalton arrived at the conclusion that matter should be atomic, it will be well to fix on the exemplification afforded by some particular combining series. For this purpose none is more convenient than the one furnished by the oxygen compounds with nitrogen, of which there are five, as displayed by the accompanying table.*

	WEIGHT.		RATIO.	
	N.	O.	N.	O.
Protoxide of nitrogen . .	14	8	1	1
Binoxide of nitrogen . .	14	16	1	2
Hyponitrous acid	14	24	1	3
Nitrous acid	14	32	1	4
Nitric acid	14	40	1	5

A glance at this table will show the justice of the remark of M. Dumas,—that, granting matter to be

* It would nevertheless be unjust to the late Mr. Higgins, of Dublin, were the circumstance not mentioned that he had entered upon certain trains of reasoning which, if persevered in, must have conducted him to the discovery of Dalton. Mr. Higgins, in his comparative views of the phlogistic and anti-phlogistic theories, published in 1789, states, p. 36 and 37, that in volatile vitriolic acid, a single ultimate particle of sulphur is intimately united only to a single particle of dephlogisticated air; and thus, in perfect vitriolic acid, every single particle of sulphur is united to two of dephlogisticated air, being the quantity to saturation. A similar train of reasoning is applied by Mr. Higgins to the constitution of water, and the compounds of nitrogen and oxygen.

atomic, it must necessarily combine as it is found to do. If, for instance, the ultimate molecule of nitrogen should weigh 14, that of oxygen weighing 8, then it should necessarily follow that, in a series of oxygen combinations with nitrogen, in which the amount of nitrogen remains fixed, the ratio of progressive increment of oxygen should be a ratio having for its arithmetical basis the number eight, of which every subsequent term should be a multiple by a whole number.

When we come to examine the volume as well as the weight of bodies entering into chemical combination, a definite order of progression will again be recognisable. Taking for instance the same five compounds of nitrogen and oxygen, and expressing by a diagram the rates of their combining volumes, the following result will be manifested :—

	N.	O.
Protoxide of nitrogen .	□	▭
Binoxide of nitrogen .	□	□
Hyponitrous acid . .	□	□ ▭
Nitrous acid	□	□ □
Nitric acid	□	□ □ ▭

From the inspection of this diagram it will be evident ; a ratio of measure as well as a ratio of weight its between the combinations of oxygen and nitrogen ; . the same remark applies to every combining series, elements of which can be obtained in the gaseous m, and their volumes estimated.

It is evident that no indication afforded by the enomena of definite chemical combination, can ever 'e us more than the *ratio* subsisting between the ight, and occasionally between the measure of *bodies* tering into chemical union. Granting, for the sake argument, the postulate of the existence of atoms, ey must be so exceedingly small that no one can hope see them, or to have any other direct manifestation of eir presence ; hence the ratio only of their combining eight, and combining volumes, is all that phenomena composition and decomposition will enable us to cognise. But the ratio subsisting between thê dimen-ns of atoms involves some curious points of specu-ion. Referring to the preceding table, in which e ratio of combining volumes between compounds of ygen and nitrogen is indicated, it will be observed at the assumed atomic size of oxygen is set down half that of nitrogen ; accordingly, starting with e assumption of the atomic size of oxygen being lf that of nitrogen, the first member in the series of mpounds of oxygen and nitrogen is considered to a combination of atom to atom ; but it would not be consistent with *rational* hypothesis, so far as facts of

of the metal being formed, and the acid of the salts evolved. It decomposes the solution of the tribasic acetate of lead; the peroxide of that metal and the ordinary acetate being formed. It rapidly converts the protosalts of iron and tin into persalts. It destroys many hydrogenated gaseous compounds: the combinations of hydrogen with sulphur, selenium, phosphorus, iodine, arsenic, and antimony are thus affected. It appears to unite chemically with olefiant gas in the manner of chlorine. It instantly transforms the sulphurous and nitrous acids into the sulphuric and nitric acids, and the sulphites and nitrites into sulphates and nitrates. It changes many metallic sulphurets (as those of lead and copper) into sulphates. It decomposes many iodides in their solid and dissolved state. By its continued action, iodide of potassium becomes converted into iodate of potassa. It changes both the crystallised and dissolved yellow prussiate of potassa into the red salt, potash being evolved. It produces oxidising effects upon most organic compounds, causing a variety of chemical changes; thus, guaiacum is turned blue by it. From the above enumeration it would appear that ozone is a most ready and powerful oxidiser, and in a great number of cases acts like Thenard's peroxide of hydrogen, or chlorine, or bromine.

When an ozonised atmosphere is made as dry as possible, and then sent through a red-hot tube, the ozone disappears, being converted apparently into ordinary oxygen, and no water or any other result is produced.

addition to the methods indicated, there are yet other
ans of obtaining ozone. Thus, for instance, if into
ar filled with ether vapour, a heated wire or glass rod
immersed, ozone is produced, and will be immediately
idenced by the test. Moreover, when solutions of
rtaric acid, certain oils, turpentine, &c., are exposed
air and light for a considerable time, they manifest a
owerful bleaching effect, attributed by M. Schönbein
the generation of ozone.

The experiments of the philosopher in reference to
il of turpentine, are especially interesting. He found
at when totally deprived of oxygen, which he accom-
lished by digestion with protosulphate of iron, it no
nger possessed any bleaching effect, and its smell was
ltered, presenting great analogy in this respect to oil
f peppermint; gradually, after exposure to heat and
ght, under occasional agitation, oxygen became
osorbed, and the bleaching effect was renewed. Cer-
in experiments of M. Osann,* will, if confirmed, still
rther disturb our pre-existing ideas of chemical combi-
ation, and increase our interest for ozone. He believes
at in certain cases when this allotropic condition of
osphorus by combination assumes a solid form, it does
t combine in the ratio of 8, on the hydrogen scale of
ity—the ratio of oxygen, but in the ratio of only 6.
e property of this conclusion, however, has been
nied by M. Schönbein, on grounds which appear

* Osann Ueber Ozen-Sauerstoff, Erdmann's Journ., Band 53
ft I., Seite 51.

synonymous with *positive pole*, and *cathode* with *negativ pole.*

The introduction of these terms suggests the propriety of others, as will be recognised. Thus, to say that oxygen is electro-negative, or that hydrogen is electro-positive, is so far inconsistent with the correctness of philosophic diction that it reposes on two assumptions; firstly, that there exist, mutually attractive and repulsive terminals (poles); secondly that the liberation of hydrogen at one of these terminals and oxygen at the other is the result of attraction.

That they are thus attracted is a position which has never been demonstrated; that they go there is merely the expression of a fact. Hence Mr. Faraday terms them, and all bodies which act similarly in regard of voltaic electricity, *ions*, from ιεμι, I go. Hence arise the compound words, an-ions and cat-ions, or, for euphony, cath-ions. A cathion is that which goes to the cathode (or negative pole); an anion that which goes to the anode (or positive pole). Hence, oxygen is an anion, hydrogen a cathion. The whole operation of a voltaio decomposition similar to that of water, is termed electrolysis, from ελεκτρον and λυω, loosen.*

---

* For details of the chemical powers of the voltaic pil the reader may consult—

1. A paper by Messrs. Nicholson and Carlisle in the Ph Mag. for 1800, and Nicholson's Journal, 4to., iv. 183.

2. Hisinger and Berzeliu[illegible] Journal, i. 115).

Process V.—By distilling the salt chlorate of potash retort of hard glass.

Theory of the Process.—Chlorate of potash is com-ed of chloric acid and potash. Chloric acid con-ns 5 equivalents of oxygen to 1 of chlorine, potash equivalent of oxygen to 1 of potassium. Therefore, ery equivalent of chlorate of potash contains 6 uivalents of oxygen, all of which are given off by the plication of heat, chloride of potassium remaining in le retort.

The rationale of the process is thus exemplified by diagram :—

Chlorate of otash = 124	1 Chloric acid = 76 . .	5 Oxygen = 40 . .	6 Oxygen = 48.
		1 Chlorine = 36 . .	
	1 Potash = 48 . .	1 Oxygen = 8 . .	
		1 Potassium = 40 . .	1 Chloride of Potassium = 76.

Process VI.—By heating in a glass retort a mix-ure of equal parts chlorate of potash and black ide of copper, or binoxide (*black oxide*; *peroxide*) of anganese.

Theory of the Process.—Similar to the above, only heat necessary to effect decomposition of the

3. Davy on some chemical agencies of electricity. Phil. ns., 1807.

4. Faraday's experimental researches on electricity. Phil. ns., 1832, *et seq.*

TABLE OF THE FUSIBILITY OF METALS.*

Class	Metal	Fahrenheit.	Authority
Fusible below a red heat.	Mercury	—39°	Different Chemists.
	Potassium	136°	Gay, Lussac, and Thenard.
	Sodium	190°	Gay, Lussac, and Thenard.
	Tin	442°	Crichton.
	Bismuth	497°	Crichton.
	Lead	612°	Crichton.
	Tellurium, rather less fusible than lead		Klaproth.
	Arsenic, undetermined.		
	Zinc	773°	Daniell.
	Antimony, little below redness.		
	Cadmium	442°	Stromeyer.
Infusible below a red heat.	Silver	1873°	Daniell.
	Copper	1996°	Daniell.
	Gold	2016°	Daniell.
	Cobalt, rather less fusible than iron.		
	Iron cast.		
	Iron malleable	require the highest heat of a smith's forge.	
	Manganese	require the highest heat of a smith's forge.	
	Nickel, nearly the same as cobalt.		
	Palladium.		
	Molybdenum	Almost infusible, and not to be procured in buttons by the heat of a smith's forge, but fusible before the oxy-hydrogen blowpipe.	
	Uranium		
	Tungsten		
	Chromium		
	Titanium		
	Cerium		
	Osmium		
	Iridium		
	Rhodium		
	Platinum		
	Columbium		

* Turner

TABLE OF THE DUCTILE METALS ARRANGED IN THE ORDER OF THEIR DUCTILITY.

Gold.	Iron.	Tin.	Palladium.
Silver.	Copper.	Lead.	Cadmium.
Platinum.	Zinc.	Nickel.	

TABLE OF THE ACTION OF SULPHURETTED HYDROGEN (HYDROSULPHURIC ACID) AND HYDROSULPHURET OF AMMONIA ON THE CALCIGENOUS * METALS. (BRANDE.)

Metal.	Solution.	Sulphuretted hydrogen.	Hydrosulphuret of ammonia.
Manganese	Neutral protochloride . .	No precipitate .	Copious ochre yellow.
Iron . .	Neutral protosulphate . .	Blackish, & small in quantity .	Black and abundant.
Ditto . .	Perchloride . .	Abundant black .	Black.
Zinc . .	Chloride . . .	A little opalescent then milky . .	Straw-colour and copious.
Tin. . .	Acid protochloride	Brown . . .	Deep orange.

* Metals admit of division into three primary classes, viz., kaligenous, terrigenous, and calcigenous—the distinction between which is founded upon the different qualities of metallic oxide. Metals, which by combination with oxygen form alkalies, are termed kaligenous, and are potassium, sodium, and lithium. Those which by combination with the same form earths, are termed terrigenous, and are calcium, forming lime; barium, and strontium, forming respectively baryta and strontia. Magnesium, aluminium, yttrium, thorium, glucinium, cerium, lanthanium, didymium, zirconium; all the remaining metals, form neither alkalis nor earths by union with oxygen, but generate substances which early chemical authors termed *calces*. Hence the term calcigenous. Calcigenous metals are alone affected by hydrosulphuric acid and *hydrosulphate* of ammonia.

At a temperature of 60° F. water absorbs two volumes of chlorine. The solution has a specific gravity of 1·008, is of a pale yellow colour, has an astringent nauseous taste, and, if aqueous moisture be present, a bleaching agency. Chlorine is not a combustible, but a supporter of combustion; the results of such combustion presenting a general analogy to the results of combination with oxygen: viz., some are acid, as hydrochloric, hydrobromic, hydriodic acids,—whilst others are not acid, and are simply termed chlorides; as chloride of magnesium, of sodium, of calcium, &c. The presence of free chlorine may be recognised by its colour, odour, and bleaching quality; by its capability of dissolving leaf gold; and by its action on soluble salts of silver, with which it occasions a white curdy precipitate, insoluble in nitric acid, but soluble in ammonia.

BROMINE—ETYMOLOGY, HISTORY, NATURAL HISTORY, PREPARATION AND PROPERTIES.

ETYMOLOGY.—Bromine, from βρωμος, *graveolentia*, on account of its peculiar smell.

HISTORY.—Bromine was discovered by M. Balard, of Montpelier, and first described in the *Annales de Chim. et Physique*, for August 1826. It was originally obtained from the uncrystallisable residue of sea water *after evaporation*, but the chief sources of it at this time, are the saline springs of Theodorshall, near Kreutznach, Germany.

NATURAL HISTORY. — Bromine is extensively, though sparingly, diffused throughout the whole ocean and certain mineral springs; always in the state of combination.

PREPARATION. — The method of preparation, or extracting bromine from its combinations, whether in the ocean or in mineral springs, is as follows :—Having evaporated the bromine-containing liquid, and having thus effected the separation of readily crystallisable bodies, chlorine is transmitted through the bittern, or fluid which remains. Chlorine having the power to effect the decomposition of bromides, bromine is set free, and remains dispersed through the fluid. Ether is now added, in which fluid bromine is soluble, hence an ethereal solution of bromine floats on the surface. All that now remains consists in effecting the separation of bromine and ether. Distillation cannot be had recourse to, since both chlorine and bromine are exceedingly volatile; therefore a somewhat complex mode of procedure is rendered necessary. The ethereal solution being saturated with potash, two bromine salts are formed; namely, bromide of potassium, and bromate of potash. These two salts being collected and exposed to a red heat, bromide of potassium alone remains. This salt being mixed with binoxide of manganese and sulphuric *acid, yields* bromine by distillation.

the two preceding simple bodies. The smell, t presents an analogy, and so do the chemical relations the three. Nor is this all: just as there is a progress relation as to cohesion, so when we come to exam the combining powers of the three, as indicated their respective equivalents or atomic weights, the sa mutual relation will be rendered evident. This circu stance has been made the basis of some beautiful spe lations by M. Dumas—speculations which have scarc yet assumed the consistence of a theory, and which only at the present time to be ranged amongst t poetic day-dreams of a philosopher:—to be regar as some of the poetic illuminations of the men horizon, which possibly may be the harbinger of a n law.

Regarding chlorine, bromine, and iodine as triad, it will be seen, as we have observed, that betw the first and the last there is recognisable a well-marl progression of qualities. Thus: chlorine is a gas un ordinary temperatures and pressures; bromine a fl iodine a solid: in this manner displaying a prog sion in the difference of cohesive force. Aga chlorine is yellow, bromine red, iodine black, or vapour a reddish violet. Here we have a chrom progression:—and, strange to say, if we refer to atomic or equivalent weights of the three, a num progression will be observable. Thus, the ato weight of chlorine is 35' of bromine 80, and of iod

25*; and now, if the atomic weights of chlorine and iodine be added together and divided by two, the result will be the atomic number for bromine.

TABLE SHOWING THE NUMERICAL RELATION BETWEEN THE ATOMIC WEIGHTS OF CERTAIN TRIADS.

Sum	Weight	Element	Mean
47	40	Potassium	
		Sodium	24
	7	Lithium	
89	20	Calcium	
		Strontium	44
	69	Barium	
80	16	Sulphur	
		Selenium	40
	64	Tellurium	

COMPARISON BETWEEN THE PROPERTIES OF CHLORINE, BROMINE, AND IODINE.

Element	State	Colour	Weight	Sum
Chlorine . .	Gaseous . .	Yellow . .	36	162
Bromine . .	Fluid . . .	Red . . .	78	
Iodine . .	Solid . . .	Purple . .	126	

Thus, we have here one of the many scientific developments of late origin, which tend to lead us back into speculations analogous with those of the alchemists. Already have we seen that it is possible for one body to assume, without combination, two distinct phases of manifestation; therefore such of the so-called elements as are subject to allotropism, are not the unchanging entities they were once assumed to be; and now we

* The atomic weights of these and many other chemical substances slightly vary according to different experimenters. The numbers here given are adopted by M. Dumas.

be thus expressed:—when three bodies having qualities precisely similar, though not identical, are arranged in succession of their chemical powers, there will be also a successive arrangement of mathematical powers, indicated by the respective atomic numbers of the substances; and amenable to every mathematical law.

"That this symmetry of chemical with mathematical function points to the possibility of transmutation is unquestionable—yet not transmutation in the sense of the old alchemical philosophy. Chemists see no manifestations of a tendency of being able to convert lead into silver, or silver into gold. These metals are not chemically conformable. One cannot take place of another by substitution. They do not form an isomeric group. The probability is that our first successful transmutation as regards the metals will effect the change of physical state merely without touching chemical composition; thus, already we have carbon, which, as the diamond and as charcoal, manifests two widely different states. Sulphur also assumes two forms, as also does phosphorus. Then, why not a metal? This sort of effect M. Dumas suggests will be amongst our first triumphs in the way of transmutation.

"In conclusion, M. Dumas cited in support of his theory the fact that bodies of conformable qualities were generally found in union or proximity, whether as presented by nature or formed by the agency of man.

"Thus, with iron there is associated manganese. Where nickel exists cobalt is not far off;—and in the organic kingdom, when man elaborates alcohol, there are simultaneously formed small amounts of ethereal bodies. Wherever viewed, chemistry is full of the startling coincidences now introduced to our notice for the first time by M. Dumas; and, whatever may be the difference of opinion as to the speculative notions of the philosopher, there can be no doubt that he has opened a wide store of chemical treasure."]

INTRODUCTION TO LECTURE III.

YDROGEN—ITS SYNONYMES AND ETYMOLOGY, HISTORY, NATURAL HISTORY, PREPARATION, AND QUALITIES.

SYNONYMES.—Hydrogen, ὑδωρ, water, γεννειν, to 'enerate. (Lavoisier.)

Inflammable air. (Previous chemists.)

Phlogiston.

HISTORY.—Hydrogen was first examined in a pure tate by Mr. Cavendish in 1766,* before which time it ad been confounded with several of its compounds, under the name of inflammable air.

NATURAL HISTORY.—Hydrogen exists largely dif'used in both the inorganic and organic kingdoms. In he *inorganic kingdom* it is a constituent of various acids n combination—as the hydrochloric, hydrobromic, hydriodic, acids. It is also a constituent of liquids—as water and naphtha; of certain solids—as sal-ammoniac and sulphate of ammonia.

In the *organic kingdom* hydrogen is found largely

* Phil. Trans. vol. lvi. 144.

entering into the composition of animals and vegetables, chiefly in the form of water and ammonia.

PREPARATION.—*Process I.*—By the electrolysis of water, each resulting gas being separately collected. This process liberates hydrogen in the condition of absolute purity.

Process II.—By adding a mixture of one part oil of vitriol by measure, and about four or six by measure of water, to comminuted zinc or iron.

Theory of the Process.—Neither zinc nor iron are capable of uniting directly with sulphuric acid: oxide of zinc and of iron, however, combine readily. Thus, a decomposition of water is determined. Oxygen of water uniting with zinc or iron, forms an oxide of these metals respectively, hydrogen being developed. In a diagrammatic representation the decomposition may be thus expressed:—

1 Water = 9 { 1 Hydrogen = 1		——— △ is evolved
{ 1 Oxygen = 8 }	} 1 Oxide of Zinc = 42—	
1 Zinc	= 34 }	}
1 Sulphuric Acid . . .	= 40	} 1 Sulphate of Oxide of Zinc = 82

Process III.—By transmitting aqueous vapour through a gun-barrel or other tube of iron heated to redness.

Theory of the Process.—The water being decomposed, its oxygen unites with the iron of the tube,—and *hydrogen* is evolved.

3.—By neither process—II. nor III.—can hydro- be obtained in a complete state of purity, in either eing contaminated with a carbonaceous result. ps, also, when zinc is used with a little of that in solution. The latter may, however, be se- ed by transmission through solution of potash.

Process IV.—By passing up into a tube inverted, filled with water, a piece of potassium or sodium.

Theory of the Process.—The potassium or sodium ting with oxygen to form potash or soda, hydrogen liberated.

PROPERTIES.—Permanently uncondensable and aëri- orm—neither taste nor smell when pure. Slightly oluble in water (108 cubic inches of water dissolve at 60° F. about 1·5 inches of the gas). May be respired once or twice with impunity, but if continued is fatal. s a combustible, but does not support combustion. Is neither acid nor alkaline,—and the sole result of its combination with oxygen is water. The lightest pon- derable body in nature, being sixteen times lighter han oxygen, and lighter than atmospheric air, in the atio of 1 to 0·0696, which latter number is, therefore, aid to be the specific gravity of hydrogen. At the emperature of 32°, 100 cubic inches of hydrogen weigh 2·22756 grains. Hydrogen is usually regarded s the unit of atomic weight, and the size of its com- bining volume is double that of oxygen.

LECTURE III.

HYDROGEN.

MATERIALS AND APPARATUS REQUIRED FOR ILLUSTRATING THE LECTURE.

Granulated zinc.

Sulphuric acid.

Water.

Taper adapted to copper wire.

Mixture of two volumes hydrogen and one oxygen in a jar standing on the pneumatic trough.

Bag and jet for blowing soap-bubbles with mixed gases.

Cavendish's eudiometer, electrical machine and Leyden jar.

Candle and oil lamp for demonstrating that water results from their combustion.

Voltaic battery and tubes for collecting oxygen and hydrogen from water decomposed.

Platinum wire to be ignited by the voltaic battery.

Apparatus for passing steam over red-hot iron and developing hydrogen.

Inverted tube filled with water, for the purpose of generating hydrogen by contact with sodium.

Spongy platinum.

Hydrogen and chlorine in tubes to be exposed to sun-light.

Bromine and hydrogen and hot wire.

Chlorine and hydrogen ready to be exploded in Cavendish's eudiometer.

Hydrochlorate of ammonia for the evolution of hydrochloric acid by [illegible] with sulphuric acid.

Hydrochloric acid developed from salt—its properties to be investigated.

[illegible] and into an atmosphere of a little chlorine to show [illegible].

I consider it fortunate, in opening this lecture "On Hydrogen," that the arrangement of elementary bodies as announced in the prospectus, enables me to take up hydrogen at the most convenient d for bringing prominently before your notice, by comparison with the elements already con-ed, some of its most remarkable points. You aware that it is not my object in delivering this rse of lectures to present you with the complete ory of the relations of the non-metallic elements, , merely to illustrate certain leading points in nection with them.

I therefore shall say nothing concerning the history hydrogen, but shall at once bring before your tice the methods of preparing this element.

Without going into details of manipulation and pparatus, let me at once proceed, in a simple way, show how hydrogen may be produced.

It is ordinarily evolved by adding either zinc or iron o a mixture of oil of vitriol and water.

To illustrate this process I take the metal zinc, ecause not only is it nearest at hand, but because he result of its action, when brought into contact rith the acid mixture, is that to which my table efers :—

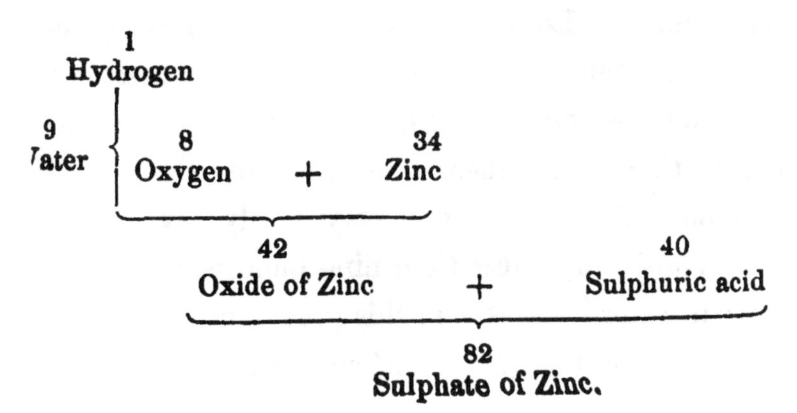

their admixed parts the power of whole thunder-storms, should wait indefinitely until some cause be applied, and then furiously rush into combination and form the bland, unirritating liquid, water, is to me, I confess, a phenomenon which continually awakens new feelings of wonder as often as I view it.

And now let us ponder for an instant on the leading qualities—*the points*—of water. Let us consider how widely it is distributed throughout nature—how numerous its functions—how tremendous its operations—and yet how mild, how bland, how seemingly powerless this wonderful liquid is. Let us view it in relation to the structure of living beings, and reflect how intimately it seems connected with vitality. Not only does it bathe the most delicate tissues and organs with impunity, but it enters largely into the composition of all organised forms. No structure of corporeal vitality is without it as an essential element. Water constitutes at least nine-tenths by weight of our own bodies, entering even into the very bones; yet this is but a trifling fraction of the amount of water entering into the structure of certain lower animals. Look at those delicate sea beings, the medusæ, and reflect on the vast amount of water which their structures contain! Pellucid almost as the ocean in which they dwell, these creatures float about in the full vigour of life; yet one may safely say that the medusæ consist of no less than nine hundred and ninety-nine parts water! Water in this great amount pervades their whole economy. Without much violence to age, we might call them living forms of water! iew these same medusæ taken from the ocean and ered on the beach, exposed to the influence of the sun air—their aqueous portions gone—what are the usæ then! Shadows almost—a substance barely— merest shred of filament and membrane!

The contemplation of this circumstance is the more nge when we consider that water, whilst in the nic structure of living beings, is continually underg changes, not so powerful in their manifestation, ough analogous in their result to the explosion which have seen—continually yielding up its elements to duce other forms of combination.

Water, you are aware, was considered to be an elent by the ancients—an opinion which has been thought be unreasonable and ridiculous by some; but for me, onfess my inability to see how the ancients, with the ount of evidence at their disposal, could have arrived any other conclusion. Viewed in its relation to the iverse—to its great natural manifestations—to the ge range of substances into which it enters—to the nifold purposes it subserves; and, more than all, ewed in relation to its intimate connection with living rms—water does seem to present to our minds the leadg qualities of an element; and it is only by the l of chemical analysis that we prove the idea to be correct.

When oxygen and hydrogen unite, the result—the le result—is water, as I have remarked, but not de-

[illegible]. Let us therefore now proceed to demonstrate this fact.

This strong glass vessel, with stopper attached, through which are made to pass two wires of platinum, is so contrived that the air which it contains may be removed by means of the air-pump, and a vacuum, or at least an approach to a vacuum, formed. Screwing this vessel, now, after its air has been exhausted, upon this jar of mixed oxygen and hydrogen in the ratio of one to two, and turning the stopcock, the water rises in the jar, as you observe—a rise which is indicative of the filling of the vessel with mixed gas. This being done, I now cause an electric spark to traverse the gaseous space between the two platinum ends, and the whole vessel becomes pervaded with light: the two gases have combined, the thick glass vessel, dry at first, is bedewed with moisture, and the moisture is water. I now turn the stopcock, and observe, more gas rushes up; and now, passing another electric spark, we obtain another flash—another combination—another contraction of volume. I might get a third charge—a third combination—perhaps a fourth; but the continuance of this experiment would be dangerous and uncertain: dangerous because each result of combustion heats the glass and increases its tendency to rupture; uncertain, because the amount of aqueous vapour formed would interfere with the action of tho electric spark. This experiment, so conclusive as to the composition of water, we owe to Cavendish, and this thick glass apparatus bears his name.

Although this form of experiment is the most perfectly demonstrative of the composition of water, yet there are innumerable phenomena of a more simple kind which bear evidence to the same fact. Thus, for instance, if I burn a jet of hydrogen under an inverted jar the sides of the jar are speedily bedewed with a liquid, which is water; and, by continuing the experiment, water in considerable quantities may, without difficulty, be collected. Nor is a jet of hydrogen indispensable to this result; any combustible body, having hydrogen as one of its component parts, producing a similar effect. Thus, for instance, I may produce water from the combustion of a candle, just as I produced water from the combustion of hydrogen, by substituting a candle in its place; and in this way it is most interesting to see the great quantities of water thus being formed as a result of the phenomenon of combustion.*

* This copious production of water by the combustion of hydrogen constitutes one of the great objections to gas-stoves without flues. These stoves neither smoke nor generate ill odours it is true, but the water which they generate is sufficiently considerable to prove injurious to articles of furniture, and to the human constitution in certain states of health.

water, and sulphuric acid, although it liberates the hydrogen from the water, yet the source of this liberation is not directly manifest. I will now effect a similar result by employing another force—the force of voltaic electricity. These two charcoal points, which terminate each respectively a copper wire, are in connection, by means of these wires, with the ends or terminals of power of a voltaic battery: and, if any one deserved to have an apparatus stamped with his name, surely that individual was Volta. Bringing these two charcoal points into contact, you observe a vivid light is given forth: and in this light we have the evidence of the transmission of power.

I will now vary the experiment by removing the charcoal points and attaching in their place two sheets of the metal platinum. These two sheets of platinum I dip into a little water slightly acidulated; but the acid has nothing to do with the final result. Observe the condition into which the water is thrown, being pervaded with bubbles throughout, and presenting an appearance very much resembling that which occurred when a mixture of water and sulphuric acid was added to the metal zinc. The sheets of platinum are not brought into contact, you will remark, but are at some distance from each other: yet the effect to which I directed your attention takes place.

In this result we have a proof that the water transmits a certain power, the same which you just now saw evidenced in the manifestation of heat, but now under the manifestation of chemical decomposition. By slight variation of the experiment we can succeed i causing this power to manifest itself under the aspect o heat and of decomposition at the same time. For thi purpose I substitute a wire of platinum for the copper wire just now used; and, observe, the platinum wire becomes red hot, whilst the water is driven into gas bubbles, as it was before.

I wish now to draw your attention to the fact that the gas bubbles in question are not evolved from every part of the water alike, but merely from those portions immediately in contact with each sheet of platinum; thus showing that the electric power is, by some means, and under some invisible form, transmitted through the intervening water.

Let us now proceed to collect some of the mixed gas which is being evolved — a matter of no difficulty. We have some now in a tube, and, applying a flame, you will easily recognise the mixture of oxygen and hydrogen by the flame and the slight explosion; but by a slight modification of apparatus it is not difficult to collect these gases separately, inasmuch as hydrogen is given off exclusively

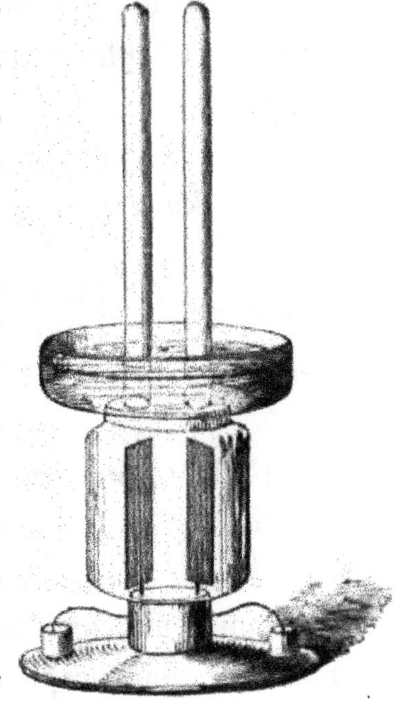

nitrogen especially characterises animal, in contradistinction to vegetable, beings; nevertheless it is found in the latter, but in small quantities. The natural orders cruciferæ and fungaceæ are amongst vegetables especially rich in nitrogen. Inasmuch as animal beings contain so much nitrogen, and vegetables so little, Berzelius has imagined that nitrogen is generated in some unknown manner by the animal functions. This idea, however, has been opposed by Liebig, who, with the majority of chemists, believes that the nitrogen existing in plants is sufficient to account for the large quantities of that element locked up in the tissues of herbivorous, no less than other animals.

PREPARATION.—(1) By the combustion of phosphorus in atmospheric air; the result of which is the formation of phosphoric acid and the liberation of nitrogen.

DIAGRAMMATIC ILLUSTRATION OF THIS DECOMPOSITION.

Atmospheric air	{ Nitrogen ————————	△ is evolved.
	{ Oxygen }	
Phosphorus . . .	}	Phosphoric Acid.

(2) By transmitting chlorine through a solution of ammonia: chlorine uniting with hydrogen to form hydrochloric acid, which combining with undecomposed ammonia, hydrochlorate of ammonia is formed, whilst nitrogen escapes.

GRAMMATIC ILLUSTRATION OF THIS DECOMPOSITION.

aonia { 1 Nitrogen = 14 ——— △ is evolved.
17 { 3 Hydrogen = 3 } 3 Hydrochloric
rine . . . . . . =108 } acid=111 } Hydrochlorate
aonia . . . . = 51 } of ammonia =162

) By agitating a liquid amalgam of lead and ry with atmospheric air for two hours or more,—esult of which is absorption of oxygen and ion of nitrogen.

) By mixing iron filings and sulphur with water, xposing the mixture to atmospheric air for the of forty-eight hours. In this, as in the preceding tmospheric oxygen is absorbed and nitrogen set free.

) By exposing muscle (flesh) to the action of acid in a retort to which heat is applied. The of this decomposition is exceedingly complex, nnot be satisfactorily explained.

OPERTIES.—Nitrogen is permanently elastic and less. It has neither taste nor smell, does not act egetable colours, nor does it whiten lime-water. s latter characteristic especially it is distinguished arbonic acid. It is neither acid, nor alkaline, supports combustion nor burns,—although its al relations impart to it some claim of being ed a combustible. Water which has been boiled s it to the extent of one and a half per cent. Its [illegible] nitric acid over the substances generally, of the inorganic kingdom I need not look farther for examples than the case afforded by pouring the acid over tin and copper respectively. In either case you observe [illegible] action take place: in either case you observe [illegible] orange coloured fumes evolved; but when tin was [illegible] a white insoluble residue remains, whereas when copper was employed there resulted a [illegible] the chemical powers of nitric acid, [illegible] which it is applied: but on these I [illegible] my object not being to give you a [illegible] of these matters.

[illegible] the compounds of nitrogen [illegible] because of its falling naturally [illegible] of my discourse, and because of its [illegible] you will see, however, by casting [illegible] at the table of compounds of nitrogen and [illegible] other substances stand before it, if we have [illegible] equivalent of oxygen with which [illegible] combined. The first compound indicated [illegible] a compound of one volume nitrogen with [illegible] of oxygen: this compound I have here [illegible] in the shape of a colourless gas. It is the [illegible] of nitrogen, or nitrous oxide — sometimes

---

[illegible] In an instant the heart shrivels [illegible] and contracts to one-third its original size. The experiment [illegible] worth seeing for once, and may be performed [illegible] by stunning the rabbit with a blow on the head [illegible] opening the thorax and displaying the heart.

lled the laughing gas, owing to the curiously exhila-ting effect it produces when breathed. This gas is enerated by the distillation of nitrate of ammonia, and fords another instance of a nitrogenous body being roduced by an indirect agency.

This gas has a sweet taste, if one may be per-mitted the application of this expression to a gas,—and ts physiological effects probably having been made amiliar to you, I will proceed to draw your attention to a circumstance illustrative of the accurately-adjusted balance of powers by which all created things are adapted, each to the fulfilment of its own peculiar end.

Not long since I remarked, that of all the associations of nitrogen with oxygen, none, so far as we knew, could discharge the functions of the atmosphere. Now here is a gas which may be breathed with impunity for a short time, but if the time of inspiration were to be prolonged, death would inevitably ensue: here, then, in one great necessity of the atmosphere, the nitrous oxide is deficient. Let us now try it in reference to combustion. Judging from the great amount of oxygen which it contains in comparison with atmospheric air, one might infer that this protoxide of nitrogen would be a better supporter of combustion than the atmosphere: this inference is not correct, absolutely. It is a better and it is a worse sup-porter of combustion than atmospheric air; but it is necessary that I explain; which explanation will best be accomplished by means of experiment. On plunging a taper already lighted into a jar of this gas, you will [illegible] with much more brillianc[illegible] [illegible] air: hence, so far as this experi[illegible] [illegible] evidence, nitrous oxide is a bett[illegible] supporter of combustion than atmospheric air. But if instead of [illegible] a fragment of phosphorus, only just ignited, be inserted—ignited to an extent that would [illegible] combustion were it allowed [illegible] atmospheric air,—then, under these conditions, the phosphorus is extinguished; a circum stance which illustrates my expression of the gas i[illegible] question being a worse supporter of combustion tha[illegible] atmospheric air. If now, varying the experiment slightly, the phosphorus be only ignited, and then plunged into th[illegible] gas, rapid and exceedingly vivid combustion supervene[illegible] —a result which shows that the power of supportin[illegible] combustion is there, but that it is only brought in[illegible] operation under the operation of certain conditions. [illegible] will be easy to recognise, therefore, for the evidence [illegible] these experiments, that nitrous oxide gas would have be[illegible] totally unfitted to stand vicariously for the atmosphere.

Next in the list of substances resulting from the com bination of nitrogen and oxygen, as shewn by the tab[illegible] comes the binoxide of nitrogen, a gaseous substance co[illegible] taining twice the amount of oxygen held in union by t[illegible] former. Like the protoxide, you will observe it colourless whilst confined in a closed vessel, as this bot[illegible] for instance; but it is different from the former, not o[illegible] in composition, but in its manifestations. Thus, i[illegible] [illegible] the stopper of a bottle containing this gas, y[illegible]

ll immediately see the portions of it which come into
ntact with the atmosphere change colour. This arises
m the formation of two new substances, hyponitrous
d nitrous acid, mixed together in varying proportions.
hese substances result from the combination of the
noxide with more oxygen; let us see, therefore, what
e result will be of mixing some of the binoxide, not
ith atmospheric air, but with pure oxygen. In this
r, standing over the pneumatic trough, is some pure
ygen, and into it I pour some of the binoxide of nitro-
n, standing in another jar: you observe the intense
nge vapours which result,—much more intense than
hen we mixed atmospheric air with the binoxide: and
u will also observe how rapidly the orange vapours
us produced are absorbed by the water. On account
' the orange-red vapours thus developed in bringing
noxide of nitrogen in contact with the atmospheric air,
oxygen, or, generally, any gas containing oxygen, the
o remaining gases become mutually tests for each other:
e appearance of red vapours on the admixture of
noxide of nitrogen with any unknown gas being indi-
tive of the presence of oxygen.

The general tendency of all the series of oxygen
mpounds with nitrogen, is towards acidity; and, in
e three last compounds mentioned, namely, hyponi-
ous, nitrous, and nitric acids, the quality of acidity is
ry highly developed. The test for general acidity is
her tincture of litmus or litmus paper. If I take
slip of the latter, and hold it in the orange fumes,

acid. A white sediment will form, which must allowed to subside: the clear solution must be decan and boiled to dryness in a glass vessel. A white m will remain, which may be fused in a platinum cruci and poured out into a clean copper dish. A transpar substance is thus obtained, consisting of phosphoric a with phosphate and a little sulphate of lime, commc known under the name of *glass of phosphorus.* yields phosphorus when distilled at a bright-red l with charcoal. (Brande.)

"Wöhler recommends, instead of the preceding, calcine ivory-black (which is a mixture of phosphat lime and charcoal) with fine quartzy sand and a li ordinary charcoal in cylinders of fire-clay at a v high temperature. Each cylinder has a bent cop tube adapted to it, one branch of which descends a vessel containing water. The efficiency of process depends upon the silica acting as an acid combining with the lime of the phosphate at a l temperature, while the liberated phosphoric acic decomposed by the carbon." (Graham.)

PROPERTIES. — Phosphorus, being one of tl substances which are subject to allotropism, assu two distinct conditions, necessary to be treated of s rately. In its ordinary state phosphorus is a taste colourless, or light-buff coloured substance, of wax consistency, semi-transparent, and at ordinary ter ratures flexible. Its specific gravity is 1·770.]

insoluble in water, but soluble in ether and in oil. At 95°, air being excluded, it melts; and at 550° it boils, yielding a vapour the density of which, according to Dumas, is 4·355. At all temperatures above 32°, phosphorus, when exposed to atmospheric air, is luminous, and a temperature of about 60° F. causes it to burst into flame. This extreme combustibility of phosphorus renders it necessary to keep the substance under water; from which it should be taken, for the purpose of experiment, with great caution, friction being carefully avoided; and whenever desirable to cut it into fragments, this operation should take place under water. Although phosphorus in its colourless or buff-coloured condition is not crystalline, yet this absence of crystalline form is only a collateral result, dependent on the ordinary method of preparation. If phosphorus be fused with about half its weight of sulphur, and suffered to cool gradually, a part of the phosphorus separates in rhombic dodecahedral crystals. (*Mitscherlich Ann. de Chem. et Phys.* vol. xxiv. p. 270.) A hot and saturated solution of phosphorus in naphtha also yields crystals of phosphorus on cooling.

Such are the leading properties of phosphorus in its ordinary or crystallisable state; but lately a new condition of phosphorus has been made known by Professor Schrötter, of Vienna, as constituting an amorphous allotropic state of the element. When common phosphorus is burned, especially in a limited supply of atmospheric air, there results around the focus of combustion a red ring of a substance formerly believed to be *oxide of phosphorus*. Professor Schrötter, however, succeeded in demonstrating that the material in question was generated under circumstances involving the exclusion of atmospheric air, and every other source of oxygenous supply. He proved, moreover, that prolonged heat, within certain limits, was the main condition necessary for the development of this red substance — oxygen being absent — and that a still higher degree of heat re-converted the red substance back into yellow phosphorus. Hence he deduced, both analytically and synthetically, that the red substance in question is merely an allotropic state of ordinary phosphorus.

Identical, however, as is the chemical composition of the two, their properties are widely different .Thus, common phosphorus inflames at a temperature slightly above 60°, whereas allotropic phosphorus only inflames at temperatures above 600, or, more properly speaking, does not inflame even then, seeing that the temperature in question effects its reconversion into ordinary phosphorus. Ordinary phosphorus is soluble in oils, naphtha, ether, bisulphuret of carbon, &c.; allotropic phosphorus is soluble in neither; hence the ordinary plan of employing bisulphuret of carbon as a means of washing out ordinary phosphorus from mixtures of the two.

Nor is the discovery of allotropic phosphorus a mere matter of theoretical interest. Already it has been applied to the manufacture of lucifer matches with nsiderable success, and a promise of ameliorating e of the most terrible diseases incidental to any anufacture. In addition to the continual danger of urning, to which the workmen engaged in the manu- cture of lucifer matches from ordinary phosphorus re exposed, there is another danger still more terrible, ecause more insidious and more difficult to be guarded against. Exposure to the fumes of phosphorus gives rise to the most frightful disorganisation of the aw bones, causing excruciating suffering, and usually terminating in death. The slightest spot of caries in the teeth of a workman suffices to become a focus of contamination for the phosphorous vapour; and even those with no such natural disorganisation occasionally suffer. Now, allotropic phosphorus gives off no fumes, neither is it combustible under the usual circumstances attendant upon the manufacture of lucifer matches; hence the advantages presented by this substance are obvious. Unfortunately, however, certain difficulties have hitherto attended its manufacture on the large scale, and limited the employment of this agent—so curious in its chemical nature, so valuable in removing the danger attendant on one of the most noxious operations in the whole list of chemical manufactures.

The leading characteristic of phosphorus is its extreme combustibility, of which you have already seen an example on the occasion of my treating of oxygen. To illustrate this property in a familiar way, let me just take a minute portion of phosphorus upon the tip of a brimstone match, and rub it — immediately you will observe the match take fire: but the phenomena and the effects of the combustion of phosphorus will be well illustrated by igniting a small portion in an open tube,—which I proceed to do.

Having placed a small fragment of phosphorus in this open tube, I apply heat and ignite it,—when, on impelling a current of air through the tube the phosphorus burns with great rapidity. The combustion having terminated, you will observe the appearance of two different residues—one being a red-coloured substance, and the other white. The latter, or white substance, is an acid compound of phosphorus with oxygen. Just this sort of result we should, *a priori*, have expected; the former was long imagined to be a combination of phosphorus with oxygen also, but in a lesser ratio than necessary to constitute an acid. Within the last few years, however, M. Schrötter, of Vienna, demonstrated that the red compound in question was merely phosphorus. No combination had taken place to form this red compound, but the phosphorus has assumed a second, or allotropic condition, just as sulphur under the operation of heat does the same.

his allotropic or amorphous phosphorus, prepared
is small quantity by our tube apparatus, is now
on the large scale, and commercially applied to
rous purposes for which common phosphorus was
rly used; and with advantages which will presently
r.

efore we can have a just appreciation of the value
otropic phosphorus, we must study the charac-
cs of this substance by comparison with phosphorus
ordinary condition. Common phosphorus is re-
bly combustible; tending to burst into flame on
plication of very slight friction or low degree of
a quality which renders it well adapted to the
se of forming lucifer matches. The quality of
lour, and its physical condition as to softness,
so points of comparison. Well, here is a lump
lotropic phosphorus, and you will observe the
nce between the two. In the first place, the
is totally different, that of the allotropic variety
dark;—then the fracture is different, that of
pic phosphorus being harshly brittle;—but the
striking difference between the two varieties of
horus is brought out by the application of friction,
heat. Common phosphorus we are obliged to keep
er, for the purpose of guarding against spontaneous
stion; allotropic phosphorus, however, may be kept
nged in atmospheric air; indeed, it may be wrapped
paper, and carried in the pocket even, with the
perfect impunity: and in this way Professor

Schrötter quite surprised us by his temerity, until we regained confidence, and became acquainted with th real qualities of the new substance. Common phosphor when rubbed takes fire; the allotropic variety, howev may be rubbed with impunity up to a certain poi after which its combustive qualities are brought o But the extreme use of allotropic phosphorus in the a will not be comprehended until you are informed the frightful ravages produced by the vapours of co mon phosphorus on those who are subjected to th influence, as is the case in manufactories of luci matches. Persons thus situated are afflicted wit disease which corrodes, ulcerates, and destroys th bones, causing the most horrible torture, and frequen death. The employment of allotropic phosphorus attended with no such calamitous results; and be capable of changing into ordinary phosphorus the application of an adequate amount of heat friction, it answers perfectly well for lucifer match and indeed most of the ordinary applications of ph phorus.

In many other respects these two conditions phosphorus present differences. Thus, for instance, power of solution in menstrua is different. Com phosphorus readily dissolves in the sulphuret of car whereas allotropic phosphorus does not. Of this I show you a ready proof, by pouring two portion sulphuret of carbon respectively on common and allotropic phosphorus. You can observe that there

ıifest difference as to the amount of solution; but we
.l have a better proof by and by. Pouring a little
ach of these solutions on two pieces of paper respec-
ly—the sulphuret of carbon, being exceedingly
atile, evaporates; when, observe—one piece of
er takes fire, owing to the amount of dry
sphorus deposited upon it, whilst the other remains
burnt.

This allotropic change of state is one of the most
ious developments of modern chemistry; and now
t our attention is directed to the subject, numerous
nifestations of this condition are recognisable. Not
least curious fact in connection with this matter is that
at is not an invariable, an indispensable condition to
veloping the allotropic condition. Thus, for instance,
I add a solution of bichloride of mercury to a solution
iodide of potassium, I may get a yellow or a red
mpound,—yet the composition of both is identical.
eat effects a similar change. On this paper is painted
red cross—the red iodide of mercury being used as a
gment. If I hold the paper on which is this red cross
er a spirit lamp flame, the redness presently changes
yellow. But I must not linger over this interesting
bject of allotropism—my time admonishes me to bring
fore your notice the curious fact of the existence of
lphur and phosphorus in the organic world—phos-
orus in large quantities, as I have already indicated;
t sulphur in amounts more small—nevertheless, its
esence in animals is universal.

[illegible] of the quantity in which they are found—[illegible] physiological reason in any case, and our [illegible] of this [illegible] has been mainly dependent on the imperfection of our tests. See what a beautiful [illegible] of demonstrating the presence of sulphur does [illegible] test afford,—and how [illegible] the distribution of sulphur in [illegible]. If we had tests of equal delicacy [illegible] constituents, we should then perhaps have [illegible] evidence of their existence.—Thus if we [illegible] analyses of organic beings we [illegible] three or four elements formerly [illegible] their invariable constituents, a long [illegible]—small in quantity, it is true—but [illegible] analyses not put there by [illegible]. Thus coming to the analysis [illegible] instead of three, it is the most [illegible] carbon, hydrogen, oxygen, and nitrogen, so [illegible] ultimate elements and probably [illegible] discovered.

[illegible] CONSTITUENTS OF WHEAT FLOUR.

Ultimate Elements.	Proximate Elements.
Carbon	Gum
Hydrogen	Sugar
Oxygen	Starch
Nitrogen	Lignine
Sulphur	Albumine
Phosphorus	Fibrine
Calcium	Caseine
Magnesium	Glutine
Silicon	Fat and Ashes

It is this extensively distributed masked or hidden
ndition of phosphorus and sulphur in the two organic
ngdoms, and the allotropic condition which each of
ese bodies can assume, to which I especially wished
direct your attention; passing over the more gross
d ordinary conditions of these elements with little
mment, as being subjects which from their promi-
nce have been extensively discussed and are well
iown.

I shall now conclude this lecture by manifesting
ie power of combination enjoyed by sulphur under
ertain conditions of heat and vaporisation—a power
hich we are very apt to forget is possessed by this
lement, simply because we do not find it, in the
rdinary range of material circumstances, placed under
onditions favourable to its exercise. We are apt, for
nstance, to regard the element sulphur as possessing
ombining properties very different in the amount of
heir vigour from those possessed by chlorine; yet I do
ot know that we are just in arriving at this con-
usion. If certain metals in a very fine state of
omminution be dropped into an atmosphere of chlo-
ne the metals take fire and burn, whereas the same
etals may be brought into contact with sulphur with-
ıt any such effect resulting. But chlorine naturally
ists in the state of gas—whereas sulphur does not.
ccordingly, if I expose copper turnings to the in-
ience of vaporised sulphur, which can be done by
eans of a very simple apparatus, you will observe

Lacertina, the *Menobranchus*, the *Axolotl*, and lastly the *Proteus Anguinus*.* All of these creatures are

---

* The Proteus Anguinus is an European animal, connected with whose existence there is a great deal of mystery. A graphic description of one of the two localities in which the animals are found, has been given by P. E. Turnbull, Esq., in his volume on Austria. Murray, 1840. It is as follows:—

"Our three guides stationed themselves at various points, and brandishing their large torches aloft, showed well this solemn cavern, with its huge pillars and dark dull waters—rendered the more interesting to the fanciful mind as being the habitation of that mysterious animal the Proteus Anguinus, which, except in one other spot (also in Carniola, near Sittich), has been found nowhere else.

"This creature appears to be bred in some much lower subterranean lake, and to be borne up into these comparatively upper regions when the waters swell. We are told that during a considerable part of the year there is no water whatever in the cave of the Magdalena. In the winter and spring it rises through crevices from below; and even then is the Proteus only occasionally discovered. It is found in the Poik (if the water to the right of the entrance be the Poik) in this cavern, but never in that of St. Catherine, nor in any other part of the river: neither is any other fish or living creature said to be found in the Poik after its first entrance below the earth. It is found also in the water which I have mentioned to the left of the entrance, supplied probably to them both from the same resorvoir or river beneath. To this water on the left we partially descended, but the drippings had rendered the soil muddy and unsafe. One of the guides, however, stationed at the bottom with his torch and hand-net, endeavoured to catch two or three of the Protei; but on his attempting to take them, they escaped under the rock.

"On our return to the inn at Adelsberg, I saw some of these creatures alive in a decanter of water, where, by changing the water every day, and without any other food, they had lived (as their owner told us) more than a twelvemonth. They were abou upplied with both lungs and gills, both so equally eveloped that naturalists vary in their opinions as to

---

even inches long, and perhaps half an inch, or somewhat less, in iameter; the form of body somewhat like an eel, but with four egs regularly jointed like quadrupeds'—the two fore-feet having hree toes, the two hinder two only; the legs about an inch long, nd the hinder ones at a great distance from the fore; colour, a greyish white; head very large, with a large broad mouth; two very small eyes, and behind the ears the gills of a fish. They have a double apparatus for breathing, and form a mixture or link of connection between the fish and the quadruped.

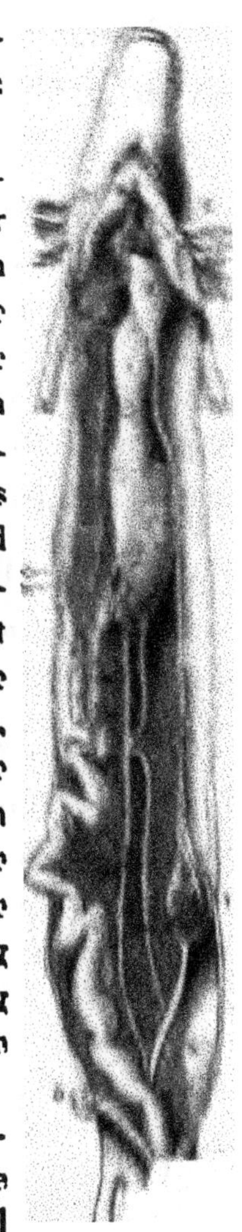

"Doomed, apparently, to live in eternal darkness in the abyss of an Illyrian mountain (for their upheaving into the caves must be considered as an exception), it might seem strange that the creature should be provided with eyes. Some imaginative writers have deemed that they had formerly a higher locality, and that amid the various convulsions of the globe, some retirement of the waters may have carried the relic of a nearly-destroyed race to its present asylum. It may be more philosophical to suppose that the small portion of light accompanying air through the crevices of the mountains, although imperceptible to our organs, may suffice to direct the course of these more delicate creatures. It is evident from the length of time that they had lived in the bottle, that the light and air of this upper world are not destructive of their vitality. Those which we saw were moving about with activity over each other, and climbing with a sort of reptile motion along the sides of the glass.

"Whether their propagation has been attempted in other places, I know not. Some were ransported to the St. Catherine Cave, and placed

The provision is very beautiful, and is as follows:—The bones of birds are hollow, and being hollow, are always filled with air: moreover, air cavities exist in various parts of the body, the gaseous element penetrating even into the muscular sheaths. Hence the lungs, not being required as vessels of capacity, but merely as instruments of chemical action, are smaller than they would have otherwise been required to exist. But there still remains to be devised a provision for filling the lungs with air, otherwise than by expansion. The muscular exercise of the animal is made subservient to this end. When at rest, and when respiration is but little required, the bird, by means of a slight muscular motion, causes sufficient air for its present necessities to enter the tiny lungs: but when the same bird is exposed to the violent exercise of flying, and when, consequently, the function of respiration is required to assume its highest grade of intensity, then the increased muscular action of the bird is made subservient to the end. Every flapping of the wings against the ribs throws a set of muscles into activity, which, by their contraction, force a largely increased supply of air into the lungs: and thus the performance of the function is accomplished as a necessary result of attendant circumstances.

Lungs of almost the same external character, thoug different in their internal conformation, are supplied t reptiles, although not for the same reason. In th serpent and lizard tribe the large, heavy, expansiv

ings of mammalia would be evidently incompatible with the conditions of their little, slender, elongated forms: but frogs have the same kind of lungs, and may be frequently seen to inflate them with air by a very simple, yet at the same time very effectual process. A frog may occasionally be seen moving inwards and outwards the sides of his face and the skin of his capacious throat. The animal is performing an act of respiration; not as we perform it, by expanding our chests: but this motion of the face and throat is precisely analogous to the motion of the leather of a pair of bellows—a motion by which the animal *forces* air into the lungs.

The chemical portion of the respiratory function chiefly involves the removal of carbon from the lungs. According to the experiments of Allen and Pepys, who were the earliest to investigate the chemistry of their function, the removal of carbon in the form of carbonic acid was the sole object; and according to them the carbonic acid corresponded exactly in proportional quantity with the amount of oxygen taken in. Later investigations, however, have proved that the explanation is not quite so simple; have proved that a portion of oxygen is absorbed, that nitrogen is sometimes absorbed, at other times given out. Still the most prominent consequence of the respiratory function in animals is evidently the removal of carbon from the blood, in the form of carbonic acid.

And now a very beautiful provision of Nature is rendered manifest to us. Carbonic acid being con-

[illegible]

[illegible]

[illegible]
[illegible]

[illegible]
[illegible]
[illegible]
[illegible]
[illegible]
[illegible]
[illegible]
[illegible]
[illegible]
[illegible]
[illegible]
[illegible] mind.

[This was the final lecture of the course of six "On the Non-Metallic Simple Elements." The theatre was very fully attended, and His Royal Highness Prince Albert honoured the audience by taking the chair.

The lecturer commenced by directing the attention of his audience to the various pieces of carbon present on the table—the diamond, coke, charcoal (animal and vegetable), plumbago, and diamonds partly converted into the black amorphous form.]

The first point in connection with the natural story of carbon, is the circumstance of its invariable lidity. This quality I shall have to allude to again in nnection with some of the most usual functions of rbon: it is one which lies at the foundation of almost ery useful application of the substance.

Notwithstanding this fixity of carbon, it is strange recognise certain chemical effects capable of being roduced by this element: thus, for instance, it has e power of absorbing larger bulks of various gases - of removing putrid smells and certain colouring atters.

Yet, (continued Mr. Faraday,) the chemical agency of arbon in an isolated state is but insignificant when ompared with the wonderful energies called into play n its various combinations. To me, viewing as I do he qualities of each element in relation to its most bvious functions in the economy of the universe— peculating on what must have happened had these ualities been different to what they are—pondering ver the all-wonderful foresight by which the harmoious balance of elemental powers was predetermined nd is maintained—this fixity of carbon, in comparison ith the volatility of its compounds, is a subject of ever-failing interest and admiration. This is a subject hich will prominently appear as I go on, and more pecially after I shall have taken up the oxy-comunds of carbon. Of these oxy-compounds there are o—the carbonic oxide and the carbonic acid gas.

CARBON WITH HYDROGEN AND OXYGEN.

Carbonic oxide	CO
Carbonic acid	CO_2
Carburetted hydrogen	CH

As regards the first, of which we have a specimen here, it is of frequent occurrence in many common instances of combustion. All who are here probably have seen a charcoal fire, and have recognised a peculiar blue lambent flame which plays upon the ignited mass. This flame is produced by the combustion of carbonic oxide. Another common instance of its occurrence is in an active limekiln: over the ignited lime there plays the same lambent flame. These instances may be sufficient, without experiment, to impress upon your minds the two leading qualities of carbonic oxide gas—its inflammability and the peculiar colour of its flame. But in this jar I have some of the gas in a pure state, and can demonstrate its combustibility and peculiarity of combustion by withdrawing the stopper and applying a light.

Of carbonic oxide little is said in comparison with the general repute of carbonic acid, yet its function in many operations are highly important: and as regards its agency on the animal economy, the experiments of Dumas have shown that it is at least a hundred times more poisonous than carbonic acid. This is a fact which was until lately unknown; carbonic acid having been considered as the more poisonous gas. However, without repeating the experiments

Dumas, they have been propounded on such evidence as I am willing to accept as perfectly conclusive. Now, this extremely poisonous nature of carbonic oxide is a fact of great significance when we regard it in connection with certain proposed plans of illumination which contemplate the employment of this gas. In itself carbonic oxide has but little illuminating power, as you have seen; but illuminating power can be given by causing it to absorb certain particles extraneous to itself, and in the end it may be made a good illuminating source. Should this plan of illumination be carried out, it will be necessary, after the experiments of M. Dumas, to watch attentively the first results; for, although the speculative chemist should be most careful not to impede, by undue expression of vague fears, the progress of any discovery, yet, on the other hand, facts so significant as those indicated by M. Dumas should not be passed by unheeded. Accordingly, I have been made aware that in France, where matters of public health are much more studied than in England—much more taken cognisance of by the Government—competent persons are anxiously watching the first effects of the new illuminative gas, considered under a sanitary point of view, and are authorised to forbid the process on the first manifestation of a result unfavourable to health.

Passing away from carbonic oxide with this short notice, we now arrive at the consideration of carbonic acid. This body, like the preceding, occurs naturally in the form of a gas, although it has been condensed

[illegible] extreme
[illegible] mixture
[illegible] little
[illegible] expose
[illegible] the
[illegible] vessel [illegible]
[illegible] have no [illegible]
[illegible]

[illegible] however [illegible]
[illegible]
[illegible] place. [illegible]
[illegible] atmosph[illegible]
[illegible] a lim[illegible]
[illegible] place [illegible]
[illegible] agitate
[illegible] similar
[illegible] which
[illegible] peculiar smel[illegible]
[illegible]

acid reaction on litmus paper, and its quality of precipitating lime-water white.]

It will be unnecessary for me to repeat the experiment, already performed on a previous occasion, of burning charcoal in oxygen gas; that experiment will have been remembered, and indeed it is sufficiently common. Less common, though now well known, is the experiment of burning the diamond in oxygen gas, and demonstrating that the result of combustion is carbonic acid. When the experiment was first performed, the result was considered extraordinary. Of course, we do not regard it as extraordinary now; but the extreme interest of the reaction shall be my apology for repeating the experiment.

[The combustion of the diamond was here effected, the gem being held by a little platinum clamp and ignited to whiteness in the oxyhydrogen flame, then plunged whilst incandescent into a jar of oxygen. Eventually the resulting gas was proved, by means of the lime-water test, to be the carbonic acid.]

Thus the evidence as to the identity of carbon with the diamond is sufficiently made out by this one experiment; nevertheless, if further evidence were required, it could be supplied by the beautiful result which I have in this glass case. Here are some diamonds which have been exposed under peculiar conditions to an ntense heat; and with the result of converting them into oke. The gems will be seen to have lost their crystalline spect—to have opened out, forming a cauliflower-like [illegible], and to have assumed the aspect of coke. These interesting specimens have been sent me from Paris, and they are most curious as furnishing us with another instance of allotropism,—that mysterious existence of identical matter in two states.

Having seen the fixity of carbon in its pure state, and the volatility of its oxygen combinations, we shall now be in a position to appreciate the nice adaptation of qualities which render it so valuable as a fuel. Had carbon not been fixed, our furnaces and fireplaces would have had no local place of action—no focus wherein their powers might have been concentrated. Had the results of combustion not been volatile, the combustive action would have been continually impeded. There can scarcely be conceived a more beautiful balance of powers designed for the accomplishment of a specific end than this; yet so familiar has the result become to us—so unnoticed by its very perfection—that an effort of chemical reasoning is required to enable us to justly appreciate this point in relation to the chemistry of carbon. The enormous quantity of ponderable, yet invisible, carbon removed in the draught of our larger fireplaces is, on its first announcement, startling; yet nothing admits of more satisfactory proof. Through an average-sized iron blast-furnace there rushes hourly no less a quantity of atmospheric air than six tons, carrying off fifty-six hundredths, or more than half a ton, of carbon in the form of carbonic acid.

Now, carbonic acid is not a supporter of combustion

as we have on many occasions seen; hence, if it had lingered in the furnace instead of being so readily dispersed, the fire could not have been maintained. This condition, as it would have been had the result of combustion been fixed, may be conveniently illustrated by a simple experiment. Taking a mass of potassium or any combustible substance, as we have seen before, I heat it on a piece of platinum-foil; yet the potassium does not take fire. Still more I urge the heat; the potassium fuses, rolls about on the platinum-foil, and becomes rapidly covered with a white crust; but it does not burn. But why does it not burn? Not certainly because of any inherent incombustibility, since, if I throw it into water, so great is its power of combustion, that it immediately bursts into flame.

Another point of deep interest in the chemistry of carbon is its strong power of illumination when incandescent. You have already remarked how trifling was the amount of light evolved from phosphorus burning in chlorine gas, or sulphur in the same gas, or sulphur in oxygen; how seemingly powerless it was, although really coexistent with highly developed chemical power. You have remarked, too, the extreme vividness of the light produced by the combustion of phosphorus in oxygen. The light developed during combustion is so far from being in any direct ratio with the accompanying heat, that the very reverse of this association may, and frequently does obtain. Thus, one of the most powerful known sources of heat is produced by burning a jet

small quantities proves fatal at once, and which l
with hydrogen constitutes that terrible—most
perhaps—of all poisons, hydrocyanic, or prus
Cyanogen cannot be prepared, at least not rea
unequivocally, by the direct combination of
elements, but is usually developed from its com
either with mercury or with silver. Its odour
of peach-blossoms, and when burned it ev
peculiar rose-coloured flame, which is very di
of this gas. Cyanogen, although a compound
buret of nitrogen), nevertheless unites with
just as though it were a simple body, and this i
its peculiarities. Thus with hydrogen it form
cyanic acid, with oxygen cyanic acid, and rep
combining tendency throughout the whole rang
compounds.

How beautiful a subject of contemplation
nished us in the calm and tranquil exercise of
laws; which a study of natural phenomena di
The world with its ponderable constituents, d
living, is made up of natural elements, e
with nicely-balanced affections, attractions, or
Elements the most diverse—of tendencies th
opposed—of powers the most varied—some s
that, to a casual observer, they would seem t
for nothing in the grand resultant of forces: s
the other hand, endowed with qualities so viole
they would seem to threaten the stability of c
yet, when scrutinised more narrowly, and ex

with relation to the parts they are destined to fulfil, all are found to be accordant with one great scheme of harmonious adaptation. The powers of not one element could be modified without destroying at once the balance of harmonies, and involving in one ruin the economy of the world!

Look at the shells of these sea mollusca: nearly one-half of their total weight is carbonic acid. Gradually it has been collected from the surrounding medium—has pervaded the systems of these delicate creatures—has circulated in their fluids—has combined with lime, and finally been deposited by their mantles in the form of a shell. Had carbonic acid been corrosive, this could not have been. Carbonic acid would have, in that case, become totally unadapted to the performance of its destined end.

And now, in bringing to its conclusion this imperfect course, I have to thank my hearers for the patience with which they have listened to me,—I have to thank his Royal Highness for the condescension shown in his visit this day. During these lectures the passing thought has often occurred that I must be bringing matters before the notice of my audience that must have been foregone knowledge, to most, if not to all. I did not set out with the intention of making these lectures a medium of systematic exposition; still less did I intend them to embrace within their scope the minute chemical relations of the non-metallic elements. I desired only to regard these elements under their more

CPSIA information can be obtained
at www.ICGtesting.com
Printed in the USA
BVHW04*1352270918
528675BV00006B/66/P